# Write a Good Undergraduate Thesis

## For Students of Biological Sciences,

## Agricultural Sciences and

## Other Related Sciences

Charles Ologidi

Dedicated to

every curious mind of the world that is ready

to learn from the wealth of knowledge of others.

## Acknowledgement

Prior to my final undergraduate year, I knew little about writing a thesis. But with the help of materials on thesis preparation, I was able to write my thesis. On that note, I want to thank everyone who in one way or the other contributed to my knowledge on how to write an undergraduate thesis.

Charles Ologidi

## Preface

An undergraduate thesis is a prerequisite for the award of a first degree. Thus students are expected to produce good theses because they are graded based on the quality of their theses. For this reason, students need materials that will equip them with the skills they need to carry out this task. This is one of such materials which provide undergraduate students of the biological sciences the know-how on writing an undergraduate thesis.

Although the focus of this book is on students of biological sciences, students of allied sciences such as agricultural science will find this book useful.

All the sections of an undergraduate thesis were fully addressed in this book.

I have discovered that undergraduates have difficulty in preparing their thesis not just because they cannot write the content of their thesis but because there are some subtle and salient points for preparing a thesis that they do not know. For this reason, I included the following into the book- some rules for writing, choosing a topic, scientific names, documentation of sources, and oral project defends/presentation of thesis.

I intend to make this material available and affordable to students while presenting them, in simplified form, with the skills they need to write a good undergraduate thesis. This is very important because sometimes the quality of your thesis is used as a benchmark for grading the quality of your undergraduate degree.

This book may not be exhaustive but you can find it useful in preparing your undergraduate thesis.

Charles Ologidi

September, 2011

# Contents

Chapter 1

## Introduction

Science, overtime, has been described as discovery. The discoveries which are made are usually communicated to the scientific community and the rest of the world through scientific papers. These papers, which are technical, have a general way in which they are prepared.

Project theses of the sciences are suppose to report the findings of a research work conducted by a student. For this reason, they can be treated as technical papers. However, the content of an undergraduate thesis is more voluminous than the content of a paper for publication.

All scientific undergraduate theses consist of the following sections: abstract (summary), introduction, literature review, results, discussion, references and appendix. Other sections include table of contents, figures (if any) and tables (if any). It is however essential that you follow the instructions given by your department/school; for this reason, there is no strict order for the presentation of undergraduate theses. Be that as it may, all sections must be presented in a chronological order such that there will be a flow of idea and a communication between the reader and the author.

Before we go further, we need to look at some rules that will be of use to you for presenting a scientific work. Some of the rules are:

1. Be concise and do not be wordy. Use short sentences where possible, because they are more direct.

2. Vary your sentences and use transitions. There should be continuity and focus.

3. Use active voice. Use passive voice only when you cannot avoid its use. Active voice is direct, precise, and clear. Passive voice is wordier than active voice.

4. Do not lie, cheat or steal. Do not copy and paste other peoples work.

5. Avoid plagiarism. You should document correctly all the sources of information for your project.

6. Be scientific and interesting. Some findings from scientific works may be cumbersome and difficult to understand but they can be written in a way a reader will find it interesting without losing the information that is being past to the across.

7. Always stay on the topic. You may want to impress your supervisor by writing so well but do not write out of point. Science deals with facts and if you do not have facts, then you should not bring in things that are irrelevant to your topic.

8. Write in a chronological manner such that there is coherence of your sentences, statements and paragraphs. You can decide to build a point from general to specific or from specific to general.

9.    At every point, you should know that you are writing to make experts and non experts to understand your topic.

10.    Use shorter versions of synonyms. For example, never utilize "utilize", always use "use."

11.    Describe past events (such as research) in past tense.

12.    Use gender-neutral language except when gender matters.

13.    Use parallel constructions. Example:

The test involves two phases:

   1. Putting the sample on the viewing stage.
   2. Raise the stage until it sits just below the probe.

Should be

The test involves two phases:

   1. Putting the sample on the viewing stage.
   2. Raising the stage until it sits just below the probe.

14. Use spelling, grammar, and style checker in your word processor. Please note that checkers do not find all errors, and they flag some "errors" that are not. But their advice is usually correct.

15. Do not worry about style or grammar in early drafts; focus on getting ideas onto paper. Then revise, revise, revise. The key to good writing is rewriting.

16. Start a new chapter on a new page.

**Note:**

Before placing pen on paper, you should have reviewed other people's works that are related to your topic. Review as many as you can because it gives you background knowledge of your topic and a whole lot to write. It gives focus to your thesis and avoids the problem of doing exactly what others have done before. However, you do not have to read everything in a scientific paper or read all the papers related to your topic in the world. You just have to scan through and study the things that are relevant and have a direct link to your topic.

## Chapter 2

### 2.0                    The Topic

A project topic must:

1.    Meet a need

2.    Be challenging and manageable within the time available

3.    Be one for which materials are available such that you will be building on the expertise of other peoples work.

4.    Allow you the opportunity for some independent work.

5.    Be one you can communicate to a variety of audiences, including non experts.

6.    Allow you to make a well-defined technical contribution and to use the project as a case study.

7.    Be one for which evaluation criteria can be established i.e., it should let you produce thesis content that can be gradable by your supervisor(s).

### 2.0.1                    Choosing a Topic

It is preferable for students to choose a topic based on their interests and future aspirations. In some departments, students are made to choose from a list of topics which saves the student the stress of formulating a project topic. In contrast, in some other departments/schools, students are made to formulate their own topic and submit it for approval.

Whatever the case may be, the selection of a project topic is better made when students consider their interests and future aspirations.

Interests vary. It is determined by area of study and quest for solutions to problems. For instance, if you know that there is no particular method for DNA extraction from mangrove species and the existing ones for extracting DNA from other plants are not very effective for these species (because of some reasons), then your topic could be "DNA Extraction from Mangrove Species"

With this as your project topic, you will be expected to give reasons why the existing methods for DNA extraction for other plant species are not effective for mangrove species. Whilst doing that, you are presenting the justification for your project. Thus, it is important you get a good topic, which is the working title for your project. It determines the content of your project thesis, from your abstract through your references.

## 2.1 Justification and Objectives

The justification for a project is the reason for the project. It is with the justification you show what problem your project is trying to address.

The objectives for a project are the expected outcomes. They are the aims/goals you expect to achieve from the project.

These two items are essential because first and foremost, they communicate clearly to your readers the problem(s) to which you are trying to proffer solution(s) to.

The justification(s) and objectives must be stated clearly. If you cannot state the justifications and objectives for your project, then you cannot do the project. It is for this reason that an application for a grant for a research work without good justification(s) and objectives cannot be approved.

For instance, if you want to extract the DNA of mangrove species, the reasons for the extraction and the expected outcomes must be clearly stated.

You may not need to have these as a separate section in your thesis. This is because it can be included in the introduction section of the thesis; in which case you will need to link the statements for your justification(s) and objectives to other statements of your introduction.

## 2.2                    Scientific Names

In order to avoid confusion and for easy classification of organisms, biologists use the binomial system of nomenclature. This system is credited to Carolus Linnaeus, for which reason he was called the father of nomenclature.

The system generally allows for all plants, animals, and bacteria to have two names. The first name is called the generic name (the name of the genus to which the organism belongs) and the second name is the specific epithet (the name of the species to which the organism belongs).

There are rules to the use of the binomial system of nomenclature. The rules include

1.    All generic names must start with a capital letter.
2.    All letters of the name except the first letter of the generic name is written in lower case.
3.    A complete name must consist of the following in this order
       •    Generic name
       •    Specific epithet
       •    Authority name
       •    Family name and
       •    Authority name of the family

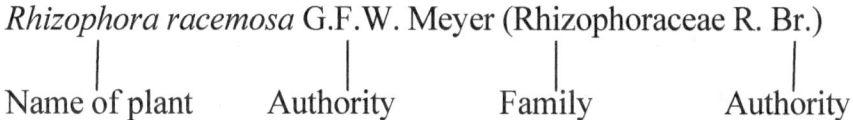

*Rhizophora racemosa* G.F.W. Meyer (Rhizophoraceae R. Br.)

Name of plant      Authority      Family      Authority

4.    The generic and specific names must be italicized (when type-written) or underlined (when hand-written).

It is unscientific for you to write an incomplete name of an organism in a project thesis, thus you are advised to adhere strictly to these rules. In addition, an incomplete name can reduce your score for your project thesis.

## 2.3                           Documentation of Sources

This is the documentation of your sources of information. The idea of documenting your sources is that anyone else should be able to retrace your steps and find the same document. Thus, your documentations must be complete and consistent. Other reasons you should document your sources are to add credibility to your work and give credit to others for their contributions.

Sources of materials are documented at the end (reference) and in the body of the thesis (bracketed citation).

Although reference and citation are used interchangeably, here, I use citation as the documentation of sources in the text and reference as a separate section i.e., as a bibliography.

The documentation of sources in the text could be either the in-text reference system or the standard footnoting/end noting system. However, the in-text reference system is commonly used.

The in-text reference system is parenthetical or bracketed citations in the text, which refer to a list of works cited at the end of the document, indicate sources of all quotations, ideas, and information. The system includes the author-date notation system and the numerical notation system.

You are allowed to use any system for documenting your sources but it is advisable to stick to one system throughout a document.

The author-date notation system puts the author of a work and the date of publication in parentheses like this (Ebiye, 2004).

The numerical notation system puts the number which the source occupies in the bibliography and the page from which the information was gotten like this (3:300).

A specific page should be cited this way (Funtari, 1997: 323), where 323 is the page number.

In the reference section, references appear in alphabetical order (by author's last name) in the works cited section at the end of the thesis. This system has variations. Most technical writing handbooks describe the APA system. (APA stands for American Psychological Association.)

**Note:** No matter what method you employ, use one system, and only one system, throughout a document.

There should be citation for all information you get from another person's work and when the authors are more than one, *et al* should follow the second name of the first author. Note the italicized "et al". It is a rule that "et al" should be italicized. For instance you quoted a statement from a paper written by Charles Ologidi and Godbless Godspower in 2012, the right citation for the statement will be (Ologidi *et al.,* 2012).

An in-text citation usually appears at the end of a sentence. Placement of the period (full stop) often creates confusion, but the rule is simple: THE CITATION IS THE LAST ITEM BEFORE THE PERIOD. Other words or punctuation marks, such as quotation marks, go before the citation.

Spacing also can create confusion. Here too the rule is simple: INSERT ONE SPACE BEFORE THE PARENTHESES OR BRACKETS.

Here is an example:

Taxon names are defined in both traditional and phylogenetic systems of nomenclature (de Queiroz, 1997).

If the material is a direct quotation of four lines or fewer, incorporate the quotation and its source into your text:

**Example**: According to H. Uhlig and R. Revie, "In recent years, annual production in the United States of all types of stainless steels, including heat resistant compositions, has reached over one million tons" (3:300).

If the material is a direct quotation of more than four lines, shorten it i.e., paraphrase.

**Example**: According to H. Uhlig and R. Revie, in recent years, annual production in the United States of all types of stainless steels, including heat-resistant compositions, has reached over one million tons. These figures represent a marked departure from the trends established earlier in this century (3:300).

If the material is an illustration, place the parenthetical citation after the caption.

**Example:** Figure 1. Pitting Potentials for Cr-Al Alloys in NaCl at 13° C. This graph demonstrates that Cr-Al alloys experience significant pitting at temperatures of 13° C (6:78).

Where should you cite references?

• Cite a quotation immediately after you use it.

• Cite ideas (things you have paraphrased) immediately after you use them.

• Cite illustrations at the end of the caption.

• If information from a source extends over more than one paragraph, cite the source in each paragraph.

If you are using the APA system and have two publications from the same author in the same year, add small letter "a" to the year of the first and small letter "b" to the year of the second. Then use the modified years in the text (Smith 2004a: 323) and in the bibliography.

How do you format bibliographic entries for online sources?

In general, you follow the format for print entries (author, title, and publication details) and add the details that are particular to online sources. These include (1) the date of access (e.g., 10 August 1999), and (2) the complete electronic address (URL).

When the online source also exists as a printed source, following this format is fairly straightforward. For sources that exist only online, it becomes a little trickier, largely because the traditional categories such as "author" and "publisher" often seem not to fit well. A good source for online citation information is OWL at Purdue: <http://owl.english.purdue.edu/handouts/research/r_docelectric.html>

Whether they appear in the text itself or in a bibliographic entry, all Web addresses should be enclosed within angle brackets < >. The acceptable standard for citing sources is evolving, but the same principles of completeness and consistency apply. If you have trouble obtaining enough detail about an online source to create a reasonably complete bibliographic citation, you probably also don't have enough information to assess the source's reliability

**Do you need a Bibliography (reference section) if you have a list of Works Cited?**

Usually a Bibliography (reference section) includes more works than those that you have cited in the document.

For example, you might include things in the Bibliography that you used as background reading, but did not cite in the text - things that another reader might want to read about the subject.

If you do not use any sources other than those on the Works Cited list, you may eliminate a separate Bibliography.

An Example: References (Bibliography)

Amjad Hameed, Salman Akbar Malik, Nayyer Iqbal, Rubina Arshad and Shafquat Farooq (2004). A Rapid (100min) Method for Isolating High Yield and Quality DNA from Leaves, Roots and Coleoptile of Wheat (*Triticum aestivum* L. ) Suitable for Apoptotic and Other Molecular Studies. *Int. J. Agric. Biol.*, **6**(2):383-387

Andrea E. Schwarzbach and Robert E. Ricklefs(2001). The Use of Molecular Data in Mangrove Plant Research. *Wetlands Ecol. Manage*, **9**:195-201

Arup Kumar Mukherjee, Laxmikanta Acharya, Pratap Chandra Panda, Trilochan Mohapatra and Premananda Das (2004). Genomic Relationship among Two Non-mangrove and Nine Mangrove Species of Indian Rhizophoraceae. *Z. Naturforsch,* **59:572**-578

The names of the authors are written first followed by the title of the paper or book. The next is the publisher's name, the volume of the paper and the pages from which you got your information.

Materials gotten from the internet:

<http://mrcgene.com/dnazol.htm>

<

http://qiagen.com/literature/qiagennews/weeklyArticle/04_03/e10/images/fig1.jpg&imgrefurl>

Note that references are written in alphabetical order from letter A-B and the first letter of each main word is written in capital letter. The publisher's name is italicized and the volume number of a journal is typed in bold.

## 2.4        Graphics

It is essential to include diagrams, illustrations, tables, and charts in the thesis. Graphic materials may be used to document your apparatus as well as to present your quantitative results. Visual materials are an effective means to convey to your reader the essence of your work. To achieve this effect, however, the graphics must be coherently integrated into your text. Most readers will probably comprehend little more than what you have explained or interpreted in the text. Therefore, be sure to discuss each figure or table in enough detail that the reader can clearly understand it.

## 2.5        Factual Statements

In science we can speculate and make opinions based on incomplete evidence. In other words, we can make guestimates based on what we know. However, some things have been established as facts and have been proven by experiments to be true. Such things are usually expressed with the use of factual statements. Factual statements are statements or assertions of verified information about something that is the case or has happened.

In technical papers, factual statements are commonly used as introductory statements of paragraphs in order to add quality to the write up.

These statements give direction to technical papers and their use indicates that the author has done a thorough review of literature by giving credit to the contributions made by others.

 As you go through this book, you will see examples of factual statements and you will appreciate the role they play in making a good thesis.

## Chapter 3

### 3.0 Summary/Abstract

The summary of your thesis gives a holistic view of your entire project such that a reader of your thesis will know the problems your project is set to address and the results you got. For this reason, your abstract should be an introduction to your work, should justify it and provide the objectives of your project. It should also provide your readers a summary of your results and materials and methods.

In this section, you will be expressing much in few words-be concise. A summary of your work should provide a complete view of your project. Thus, it should contain some aspects of your introduction, literature review, discussion, results, and materials and methods.

What you do here is simply to bring out the main points in each sections of your thesis and summarize all the points in less than a page. For this reason, it is advisable you write your summary/abstract after writing every other section of your thesis.

It is actually a prelude to the other sections hence it should give your readers all the information needed to know what your topic is all about while setting the pace for the other sections of the thesis.

For instance, a summary of the topic: **DNA Extraction from Mangrove Plants using DNAzol Protocol:**

The use of DNA data in mangrove studies has increased dramatically in the areas of investigating population structure, phylogenetic relationships and biodiversity. These areas of study make use of molecular techniques such as restriction enzyme digestion, PCR, microsatellite analysis, AFLP, RFLP, and SSR whose success depends on a DNA extraction protocol that balances time and cost with quality, yield and purity of DNA **(INTRODUCTION)**. The presence of high content of tannin, polyphenolic compounds and polysaccharides, however, limit extraction conditions from obtaining the balance. Nevertheless, presented here is a DNAzol protocol **(METHOD)** that is simple, fast and convenient for extracting DNA from three mangrove species **(JUSTIFICATION)** *Rhizophora mangle* L. (fresh and senescent leaves), *Rhizophora racemosa* G. F.W. Meyer (fresh leaves) and *Avicennia germinans* (L.)L., P **(PLANT MATERIALS)**. Extracting DNA from other mangrove species is also possible. Obtained DNA was of high quality, yield and purity. The OD260/OD280 ratio was 1.8+/-2 and the DNA bands on the gel pictures show intact and undegraded DNA **(RESULTS)**.

**Note**: you do not need to include citations here.

3.1                              Introduction

For an undergraduate thesis a page or two is enough to introduce your project.

The introduction provides the background information and the justification for your project. It prepares your readers for the information ahead and gives them reasons to read other sections of your thesis.

 An introduction is an expanded form of the abstract/summary except that it does not contain the results but the objectives of the project.

You should include some factual statements, a prelude to your materials and methods and some other things you feel should be a part of your introduction.

When writing a one-page introduction, you do not need to have subtitles for your objectives, justification and other parts of the introduction but you are to place your points in good sentences and paragraphs. Thus, include subtopics in your introduction only when necessary.

Below is an example of an introduction.

Topic: DNA Extraction from Mangrove Plants Using DNAzol Protocol.

DNA extraction protocol for plants have limitations between speed and purity in high-throughput molecular techniques such as gene mapping and marker-assisted selection programs (Xu *et al.*, 2005). **(A factual statement)**

A rapid and efficient method for extraction of DNA from mangroves is necessary for various molecular techniques used as tools for molecular biological studies of mangroves. However, the extraction of high-quality DNA from these species can be time-consuming, laborious and expensive because contaminating substance such as polyphenols, tannins and polysaccharides, which occur in high amount, interfere with the downstream enzymatic manipulations of DNA. **(A PART OF THE JUSTIFICATION)**

Mangroves are distributed circumtropically, occurring in marine intertidal zones in 123 countries and territories of the world (Kathiresan and Bingham, 2001). The Niger Delta accounts for 5% of the total mangrove area where seven species are represented and six occurring naturally. The species include *Rhizophora racemosa* G.F.W. Meyer, *Rhizophora harrisonii*, *Rhizophora mangle* L. (Rhizophoraceae); *Avicennia germinans* (L) L., P. (Avicenniaceae), and *Nypa fruticans* (Palmaea). **(A BRIEF NOTE ON THE MANGROVES OF THE NIGER DELTA).**

Mangroves are ecologically and economically important (Tomlinson, 1986), they provide ecosystem goods and services; as a result, they have been a hotspot in the fields of biodiversity, phylogeny and evolution (Duke *et al.*, 1998) **(the importance of mangrove species-making a case for the justification of the project)**. Nevertheless, no study, at the molecular level, has reported on the mangroves of the Niger Delta. In order to begin an extensive molecular biological study, it is necessary to extract DNA with an efficient method that is fast and reliable. **(JUSTICATION)**

Accordingly, in this study, DNA extraction was done with the DNAzol protocol for DNA extraction from samples collected on ice and in DNAzol( **MATERIALS AND METHODS)**; thus allowing for comparison between both methods of preservation. In addition, to be compared is DNA extraction from young and senescent leaves of *R. mangle* **(OBJECTIVES)**

From the introduction above, you can discover that an introduction explains every aspect of a topic. If the writer wishes to place the points in subtopics then we would have seen topics such as:

1.    Overview or background information
2.    Justification
3.    Objectives
4.    Procedure for Extraction.

**3.2**             Literature Review

In this section, you will be assembling information from a wide range of published literature because you will be presenting other peoples' works in relation and relevance to your project. Published literature however plays an important role in all stages of your project. For this reason, literature review is an ongoing process in which you gather the expertise and information you will need to shape, justify, execute, and document your project.

The aspects of other peoples' work you are to write about include materials and methods, abstract, results etc. You are expected to present their successes and scrutinize their faults and inadequacies whilst making a comparison between them. Thus when you mention the results of an author, you should compare it with another author's results; and do not forget that those results have to be related to your topic.

For instance when writing a literature review on DNA extraction from mangrove species of the Niger Delta and you have a paper on Genomic Relationship among Two Non-mangrove and Nine Mangrove Species of Indian Rhizophoraceae, you do not go writing about the relationship between mangrove species but you write only about how the DNA of the mangrove species were extracted.

The scrutinizing of other peoples faults and inadequacies gives you an opportunity to make a case for your project and to write a good discussion.

With your literature review, you will be looking at the past, present, and future of your topic and be asking intelligent questions that will prepare you for a critical conversation.

It is mostly convenient to make a good literature review when you list out the different aspects of your project topic and discuss them separately in subtopics. Getting the topics will not be difficult if you fully understand your project and you have adequately studied other works that are related to your topic.

Take for example the topic: DNA Extraction from Mangrove Species of the Niger Delta. The following topics can be chosen as part of the literature review:

1.     General overview of plant DNA extraction
2.     Disruption of cell wall and nuclear membrane
3.     Separation of DNA from naturally occurring cell constituents
4.     Sequestering secondary metabolites and polysaccharides
5.     DNA precipitation
6.     Optimizing extraction conditions for mangrove species
7.     The use of DNA data in mangrove plant research

What the author of this thesis would do here is to discuss the different stages of DNA extraction from disruption of cell wall to separation of the DNA from other cell constituents whilst laying emphasis on the extraction of DNA from mangrove species. The first topic will give an overview of DNA extraction from all plants. The second to the fifth will explain the order wherein DNA is extracted from plants and indeed from mangroves. The sixth will look at ways of modifying/enhancing/improving existing methods of plant DNA extraction for DNA extraction from mangrove species. Finally, the last topic will look at the uses of the DNA gotten from the extraction, which has to do with the relevance of DNA extraction from mangrove species-the relevance of the topic.

You can see the chronological order in which the topics were arranged. That is how you are expected to arrange your topics, and when you explain the topics, there should be coherence between your statements, sentences, paragraphs and topics. There should be order and focus.

3.3                     Materials and Methods

The materials are the laboratory and/or field equipment which you used for your bench work and/or field work. The samples collected are also materials and are included in this section.

The materials include, but not limited to the following: plant materials, animal materials, and chemicals, test tubes, Petri dishes, autoclave, pipette, and measuring instruments.

The methods are the procedures with which you carried out your field and/or bench work. They are the steps you employed to get your results.

It is advisable to title your methods, if possible; and present them in a step-by-step manner such that someone can easily understand the procedure(s) for your method(s). In addition, it allows for easy repetition of your work- this is one of the things that make science to be universal.

It is advisable to address your materials and methods separately whilst itemizing the materials and procedures you used in your work.

If for any reason you are unable to write all the materials and methods, you should ensure that you put down all the major materials and methods i.e., the ones that were very important to the successful completion of your bench or field work.

Depending on your project topic, you could have more than one procedure which must be stated clearly and separately from one another by titling.

Here is an example of materials and methods.

Topic: DNA Extraction from Mangrove Plants Using DNAzol Protocol

### Plant Material

Four leaf samples from three species: *Rhizophora mangle* L. (fresh and senescent), *Rhizophora racemosa* G.F.W. Meyer (Rhizophoraceae R. Br.); *Avicennia germinans* (L.)L., P.(Avicenniaceae Endl. Ex Schinzl.)- all fresh and young- were collected on ice and in DNAzol from Eagle Island, Port Harcourt in the Niger Delta.

Note the italicized names of the plants and the words that follow them. The words in front of the names of the plants are the family names and the authority names of the plants. This is very important because botanical names are incomplete without them. The "fresh, young and senescent" shows the nature of plant materials that were used for the project. This is important especially when you are doing a project in plant sciences.

*Rhizophora racemosa* G.F.W. Meyer (Rhizophoraceae R. Br.)

Name of plant     Authority          Family     Authority

You should also note the location from which the samples were collected.

## Preparation of Plant Material

30mg of each sample was homogenised in 1ml of DNAzol reagent with a hand held glass/Teflon homogeniser whose tolerance is > 0.14 mm. 8 strokes were applied.

*Note the materials that were used for the preparation of the plant material. Also take note of the quantity of the materials that were used.*

## DNA Preparation

DNAzol protocol is based on the use of a novel guanidine-detergent lysing solution which permits the selective precipitation of DNA (Chomczymski *et al.*, 1997).

The samples were homogenised in DNAzol and centrifuged. It was then followed by DNA precipitation in 100% ethanol(x2) and DNA solubilisation in 8mM NaOH.

*Note the materials that were used for the DNA preparation. Also take note of the quantity of the materials that were used.*

## Chemicals

1.  1ml DNAzol reagent

2.  0.5 ml absolute ethanol

3.  1ml 75% ethanol

4.  8mM NaOH

*Take note of the concentration and volume of the chemicals.*

Protocol – *this is the method for the extraction of the DNA. It was titled protocol and not method because it is a project on DNA extraction. You should title your methods in like manner.*

1. 30 mg of plant tissue was homogenised in 1ml of DNAzol and stored for 5min at room temperature.

   -this step lyses the cell and releases its content into solution.

2. The homogenate was transferred to a microcentrifuge tube and centrifuged for 10 min at 10, 000 x g at room temperature.

   -this step removes insoluble tissue fragments, RNA, excess polysaccharides, polyphenolic compounds and proteins.

3. The resulting viscous supernatant was transferred to a fresh tube and 0.5ml of 100% ethanol per ml of DNAzol added to the homogenate for DNA precipitation.

4. Samples were mixed by inverting 6-8 times and stored at room temperature for 2 min. The precipitated DNA was then transferred to a clean tube by spooling with a pipette.

5. The tubes were stored upright for 1min and the remaining homogenate aspirated from the bottom of the tube.

6. The DNA precipitate was washed twice with 1ml of 75% ethanol. At each wash, the DNA was suspended in ethanol by inverting the tubes 3-6 times.

7. The tubes were stored vertically for 1min allowing for the DNA to settle to the bottom of the tubes and the ethanol removed by pipetting.

8. The DNA was air-dried in an open tube for 5 seconds and dissolved in 8mM NaOH by slowly passing the pellet through a pipette tip.

9. Insoluble materials (mostly polysaccharides) were removed by centrifugation at 12, 000 x g for 10 min

10. The pH of the DNA solution was adjusted to 8.2 by 1M HEPES per 1ml of 8mM NaOH.

The extracted DNA was subjected to electrophoresis on 1% agarose gels in 0.5xTBE electrophoresis buffer at 70V for 2 hours to check the quality. Quantification of the DNA was performed in a cuvette spectrophotometer at 260nm and the purity was checked from OD260/OD280 ratio.

### 3.4 Results

The results of your project are the outcomes of your bench and/or field work. They are the data you recorded from the experiments you did and the measurements you took.

Your results can be numerical or graphical depending on your project topic. For numerical results you are advised to place them in tables which must be properly titled. The titles should include a name for the table, table number, captions, and column titles; include row titles when necessary.

Graphical results should be properly titled with names, figure numbers, and captions.

The proper titling of your results is essential because a table or figure without a name or caption means little or nothing and it makes it difficult for your readers to understand your results.

Below are results for the topic "DNA extraction from mangrove plants using DNAzol Protocol" which were presented in tabular and pictorial forms.

## Spectrophotometric Readings

| Sample no | OD$_{260}$ABS | OD$_{260}$NET | OD$_{280}$ABS | OD$_{280}$NET | 320 | RATIO | CONSTANT | DF | CONC (ng/ul |
|---|---|---|---|---|---|---|---|---|---|
| Blank | 0.000 | 0.000 | 0.000 | 0.000 | 0.000 | 0.000 | 50 | 250 | |
| 1 | 0.032 | 0.031 | 0.018 | 0.017 | 0.002 | 1.848 | 50 | 250 | 387.5 |
| 2 | 0.068 | 0.068 | 0.035 | 0.035 | 0.000 | 1.947 | 50 | 250 | 850 |
| 3 | 0.043 | 0.042 | 0.024 | 0.023 | 0.001 | 1.816 | 50 | 250 | 525 |
| 4 | 0.050 | 0.048 | 0.027 | 0.025 | 0.002 | 1.912 | 50 | 250 | 600 |
| 5 | 0.024 | 0.023 | 0.013 | 0.013 | 0.001 | 1.834 | 50 | 250 | 287.5 |
| 6 | 0.068 | 0.066 | 0.036 | 0.034 | 0.002 | 1.924 | 50 | 250 | 825 |
| 7 | 0.070 | 0.067 | 0.038 | 0.036 | 0.003 | 1.883 | 50 | 250 | 837.5 |
| 8 | 0.144 | 0.133 | 0.084 | 0.074 | 0.011 | 1.813 | 50 | 250 | 1662.5 |

Blank=Control     (no DNA, only buffer)

1-4: DNA of samples stored in DNAzol

5-8: DNA of samples stored on ice

1: *Avicennia germinans*          5: *Avicennia germinans*

2: *Rhizophora mangle*           6: *Rhizophora mangle*

   (Fresh leaves)                     (Fresh leaves)

3: *Rhizophora mangle*           7: *Rhizophora mangle*

(Senescent leaves)                  (Senescent leaves)

4: *Rhizophora racemosa*         8: *Rhizophora racemosa*

## Agarose Gel Pictures

M    -TIVE CONTROL        1   2   3   4

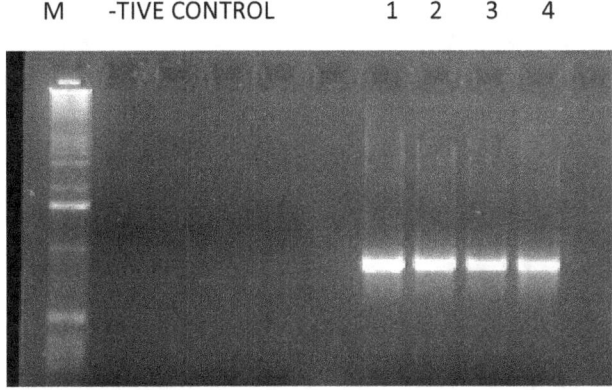

Fig 1: M = marker, Negative control = No DNA, 1-4 = DNA of samples stored in DNAzol

M    -TIVE CONTROL        5   6   7   8

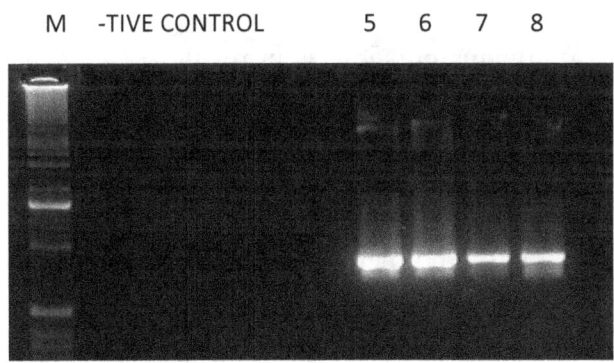

Fig 2: M = marker, Negative control = No DNA, 5-6 = DNA of samples stored on ice.

Gel electrophoresis with 1% agarose run at 70V for 2 hours

1: *Avicennia germinans*  5: *Avicennia germinans*

2: *Rhizophora mangle*  6: *Rhizophora mangle*

    (Fresh leaves)    (Fresh leaves)

3: *Rhizophora mangle*  7: *Rhizophora mangle*

(Senescent leaves)    (Senescent leaves)

4: *Rhizophora racemosa*  8: *Rhizophora racemosa*


Note: Results that are not properly titled are incomplete and do not make sense. They tell little or nothing about your project and it will be difficult for you to explain them in the discussion section.


3.5                    Discussion

This is the section where you give insights into your results.

A good literature review sets you off for a good discussion because basically what you do here is comparison. You compare your materials, methods and results with other people's that you may have written about in your literature review. Therefore, you will be making comparisons between your project and other peoples' work that are related to yours; however the comparisons are directed towards the importance and perhaps the superiority of your project. Therefore, a well reviewed literature will be of great help to you here.

In this section, you can also write about your inadequacies and the reasons behind them.

An example of discussion

Topic: DNA Extraction from Mangrove Plants using DNAzol Protocol.

**(A FACTUAL STATEMENT)** Isolation of genomic DNA ideal for a wide range of molecular biology applications like biodiversity studies is a fundamental requirement (Hague *et al.*, 2008). The protocol reported here uses the DNAzol Reagent (Invitrogen), a patent-pending DNA extraction reagent containing guanidine thiocyanate and a detergent mixture **(THE METHOD USED)**.

**(A COMPARISON BETWEEN THIS METHOD AND OTHER METHODS)** The protocol is faster than other published protocols for mangrove species (Perez, 2001; Saghai-Maroof *et al.*, 1984; Huang *et al.*, 2003; Hameed *et al.*, 2004; Mukherjee *et al.*, 2004; Parani *et al.*, 1997); the whole DNA extraction procedure took about 30 min. This is because DNAzol is a ready-to-use reagent and the incubation time was short, lasting for about 8 min.

This protocol involves less steps, microcentrifuge tubes and time. Unlike the CTAB-based published procedures for mangroves (Parani *et al.*, 1997; Perez *et al.*, 2001), phenol/chloroform/isoamyl alcohol was not used in this procedure, thus eliminating the need for a fume hood. In addition, the sequestering of insoluble tissue fragments,

RNA, excess polysaccharides, polyphenolic compounds and proteins were done at a single step. However, another step of centrifugation was used for insoluble materials (mostly polysaccharides) which were not removed earlier. Invariably, the smaller number of steps for removing contaminants reduced the hands-on time.

**(DISCUSSING THE PLANT MATERIAL)**The samples collected on ice were transferred to a freezer and stored for 2wks before DNA extraction. The samples in DNAzol were kept at room temperature. The samples on ice were effectively powdered by first freezing in liquid nitrogen.

**(DISCUSSING THE RESULTS)** The gel pictures (Figs 1 and 2) show well resolved bands, indicating that the tannin, polysaccharide and polyphenolic compounds were sequestered, unlike in Yu and Web, 2000 in which DNAzol protocol was used for *R. mangle*, *R. stylosa* and *R. apiculata*. The OD260/OD280 ratio in table 2 range from 1.813-1.947 which is within the 1.8+/- 2 spectrophotometric values for pure DNA. Thus, the DNA extracted from all samples is pure and high in quality. Likewise, the concentration ranging from 600-1662.5ug/ul is satisfactory for further analysis of mangrove DNA.

The bands of DNA of samples stored in DNAzol were more resolved than for the once stored on ice. Thus, the quality is higher. Meaning that DNA degradation in the reagent is minimal. However, the yield is generally higher with samples on ice, the highest being DNA from samples of *Rhizophora racemosa* G.F.W. Meyer. But yield for both is sufficient for further molecular analyses.

The DNA purity was okay for both samples collected on ice and in DNAzol because the OD260/OD28O ratio was 1.8+/-2. Thus, both methods of preservation produced good DNA yield, quality and purity; however, the samples in DNAzol produced higher quality.

The DNA extracted from *Rhizophora mangle* L of fresh and senescent leaves was of good quality, purity and quantity and are thus both suitable for DNA extraction from this species. Meanwhile, the samples in DNAzol produced higher quality. Unlike in other mangrove species, DNA extraction from senescent leaves of *R. mangle* L. was successful because the total organic and tannin contents decrease during senescence (Kandil *et al.*, 2004).

**(COMPARING THIS METHOD WITH ANOTHER METHOD)**The DNeasy kit has been proved useful for DNA extraction from mangrove species (Schwarzbach and Ricklefs, 2000); the cost of the kit, however, limits its use. But the DNAzol reagent used in this study is cheaper and offers more economic means of extracting DNA from mangroves.

## 3.6　　　　　Conclusion and Recommendation

The conclusion is devoted primarily to the significance and interpretation of the facts rather than to a presentation of the facts themselves, which would have been presented in earlier chapters.

Whilst the introduction presents the readers with the background knowledge and prepares them for the understanding of the other sections of the thesis, the conclusion relates to the reader in terms of the knowledge gained from the introduction through the discussion.

With that in mind, you can make a case for the importance of your project, for further studies/research, and for your limitations.

The recommendation part of it can include the call for a change of policy, for further studies, for a change of attitude etc.

An example of conclusion and recommendation

Topic: DNA Extraction from Mangrove from Plants Using DNAzol Protocol

DNA extraction is the first step to the success of any molecular technique. DNA extraction of superior quality, purity and of reasonable quantity is a prerequisite for various molecular techniques in molecular biological studies of mangroves **(FACTUAL STATEMENTS)**. As shown in the results, the DNAzol protocol is suitable for mangrove species of the Niger Delta because the procedure yields DNA of high quality, quantity, and purity. It is also fast and economical. In addition, the procedure used here offers a cheap preservation method- preservation of plant samples on ice. Because of these reasons its use is recommended for DNA extraction from mangrove species of the Niger Delta especially when hundreds of samples need to be analyzed for genetic diversity and evaluation of population structures. Although, dried and mature leaf samples were not used in this study, it is most likely going to yield good results for these samples. Thus, further studies could look at this area whenever the need arises.

## 3.6      Project Defends/Oral Presentation of Project

The grading of project works involve the grading of project theses and the grading of project defends/oral presentation of project work. Thus oral presentation of project work to fellow students and faculty members make a complete project work, with the project thesis.

There is usually a deadline to the submission of project thesis, after which students are required to defend their project work. The project defends or oral presentation of project work is to ensure that the students understand their project work and they were the ones that actually wrote the project thesis.

In most cases, project students are required to present their project work with PowerPoint slides to students and faculty members of the department. In this case, the lectures act as the jury and grade students based on different criteria such as composure, flow of ideas, and ability to communicate with the audience.

In oral presentation of your project work, you should prepare your slides well before time so that you can spot mistakes and make corrections before the day of presentation. You should include only key points about your project in the slides and the number of slides should be prepared according to the time limit allocated for students to present their project work.

The slides should not be bulky. They should contain only bullet points of which you are expected to give insights during the presentation.

You can include images only if you can explain their relevance to your project topic.

At the end of your presentation, you will be asked a few questions based on your project topic and your PowerPoint slides. Your ability to answer the questions is important because it is one of the criteria for grading your presentation.

In some cases, students are not required to present their project work before faculty members and fellow students, but are examined by a supervisor invited by the department- an external supervisor. The external supervisor grades students after going through their project work.

The external supervisor asks each student a few questions and grades them based on their ability to answer the questions.

The supervisor can ask you any question related to your project work, thus you should be prepared to answer any question.

It should be noted that some departments use only project defends whilst some others use both project defends and oral presentation of project for assessing their students' project work.

## References

Ologidi, C. G. (2011). Undergraduate Project Thesis: DNA Extraction from Mangrove Plants Using DNAzol Protocol. Unpublished.

Nyananyo, B. L. (1997). Biological Nomenclature and Species Concept. Belk Pulbishers, Belk House.

Undergraduate Thesis Manual, 2006-2007, (2006). University of Virginia. Charlottesville, Virginia 22904-4744.